Atoms and Cells

CONTENTS

Library of Congress
Cataloging-in-Publication Data

Bender, Lionel.
Atoms and cells/Lionel Bender.
p. cm. -- (Through the microscope)
Summary: Text and photographs
introduce microscopic plant and animal
life, viruses, microspores, and other
forms of life which can only be viewed
through a microscope.
ISBN 0-531-17219-8
1. Microscope and microscopy--Juvenile
literature. 2. Cells--Juvenile literature. 3.
Microorganisms--Juvenile literature. 4.
Biomolecules--Juvenile literature. [1.
Microorganisms. 2. Cells. 3. Microscope
and microscopy.] I. Title. II. Series.
QH278.B46 1990
578--dc20   89-26066 CIP AC

Printed in Belgium

Design David West
Children's Book Design
Author Lionel Bender
Editor Roger Vlitos
Researcher Cecilia Weston-Baker
Illustrated by Alex Pang

First published in
the United States in 1990 by
Gloucester Press
387 Park Avenue South
New York NY 10016

Atoms and Cells

Lionel Bender

GLOUCESTER PRESS

New York : London : Toronto: Sydney

LOOKING CLOSER

Microscopes and magnifying glasses work by using lenses and light. A lens is usually a thin, circular glass, thicker in the middle, which bends rays of light so that when you look through it an object appears enlarged. A microscope uses several lenses. It will also have a set of adjustments to give you a choice over how much you want to magnify.

When we want to view something under a microscope, it must be small enough to fit on a glass slide. This is put on the stage over the mirror and light is reflected through so that the lenses inside can magnify the view for us. But not all microscopes work this way. The greatest detail can be seen with an electron microscope which uses electron beams and electromagnets.

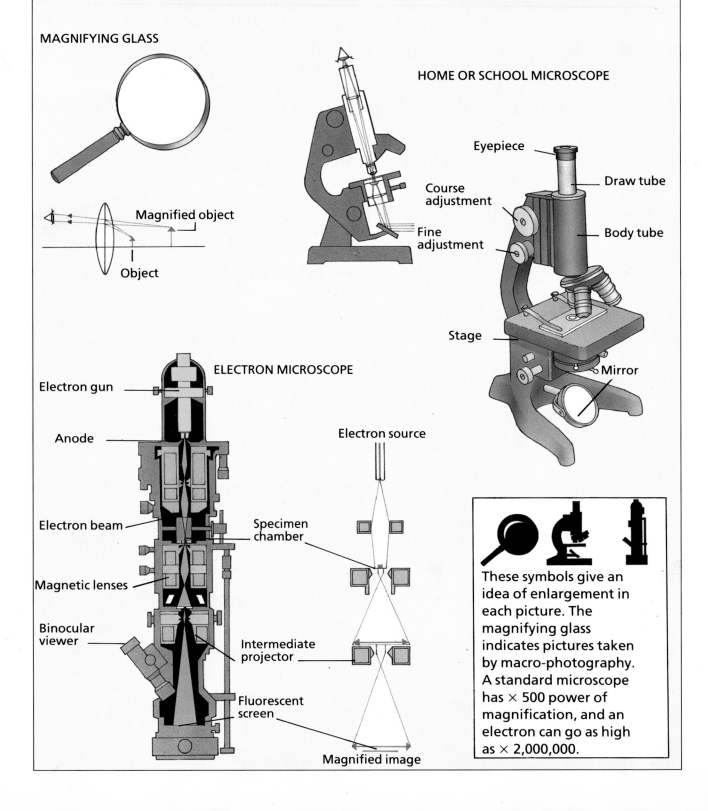

MAGNIFYING GLASS

Magnified object

Object

HOME OR SCHOOL MICROSCOPE

Eyepiece

Draw tube

Course adjustment

Body tube

Fine adjustment

Stage

Mirror

ELECTRON MICROSCOPE

Electron gun

Anode

Electron beam

Magnetic lenses

Binocular viewer

Specimen chamber

Intermediate projector

Fluorescent screen

Electron source

Magnified image

These symbols give an idea of enlargement in each picture. The magnifying glass indicates pictures taken by macro-photography. A standard microscope has × 500 power of magnification, and an electron can go as high as × 2,000,000.

INTRODUCTION

A microscope is used to study things too small to be seen with the naked eye. This book has pictures taken through microscopes, or with special magnifying lenses attached to cameras. Drawings appear alongside to help explain what the microscopes are showing us. All things, living and nonliving, are made up of particles known as atoms. We can't see atoms properly because they are so minute. But atoms combine to form molecules and, in living things, collections of certain molecules form cells. In this book we look in turn at atoms, molecules and cells.

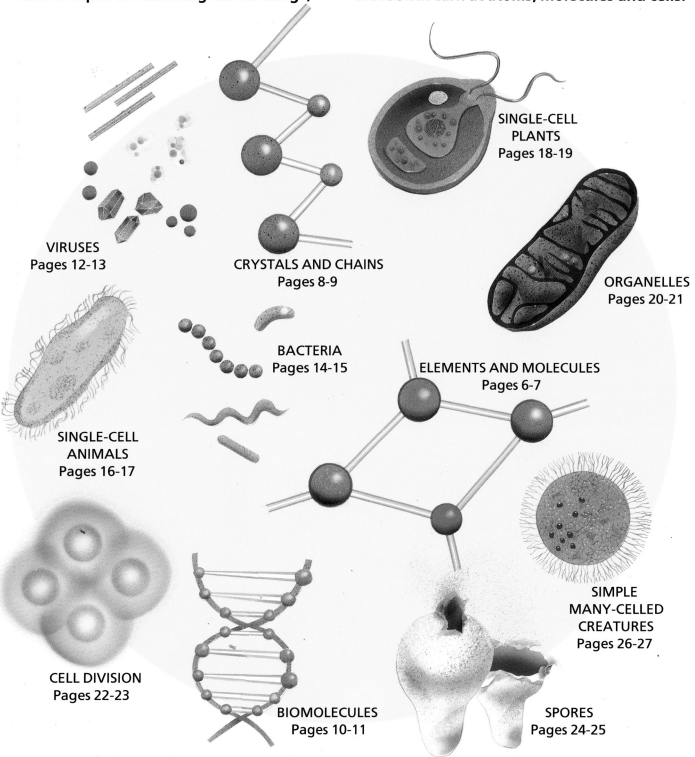

SINGLE-CELL
PLANTS
Pages 18-19

ORGANELLES
Pages 20-21

VIRUSES
Pages 12-13

CRYSTALS AND CHAINS
Pages 8-9

BACTERIA
Pages 14-15

ELEMENTS AND MOLECULES
Pages 6-7

SINGLE-CELL
ANIMALS
Pages 16-17

SIMPLE
MANY-CELLED
CREATURES
Pages 26-27

CELL DIVISION
Pages 22-23

BIOMOLECULES
Pages 10-11

SPORES
Pages 24-25

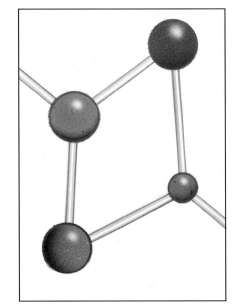

ELEMENTS AND MOLECULES

If you cut this page into smaller and smaller pieces, you would eventually get thousands of tiny fragments of paper. Under a microscope, and carefully using a very fine blade, it would be possible to cut up each of these fragments into thousands of yet smaller pieces. Finally, though, you would not be able to divide up the fragments any more. You would then have a mass of atoms. The word atom comes from the Greek, *atomos*, meaning uncuttable or indivisible. An atom is the smallest particle that can exist naturally. The pages of this book are each about 2 million atoms thick. In nature there are less than 100 different types of atom, yet these combine in endless ways to make bigger units known as molecules.

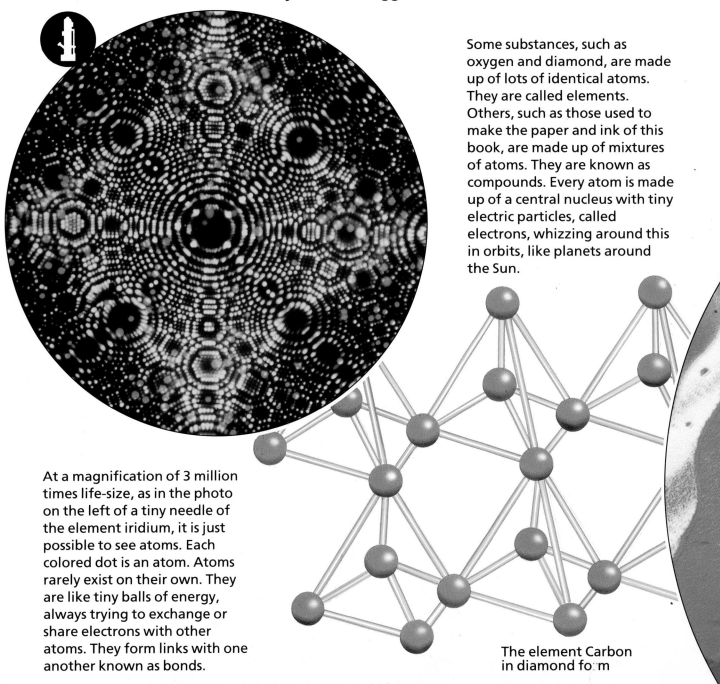

Some substances, such as oxygen and diamond, are made up of lots of identical atoms. They are called elements. Others, such as those used to make the paper and ink of this book, are made up of mixtures of atoms. They are known as compounds. Every atom is made up of a central nucleus with tiny electric particles, called electrons, whizzing around this in orbits, like planets around the Sun.

At a magnification of 3 million times life-size, as in the photo on the left of a tiny needle of the element iridium, it is just possible to see atoms. Each colored dot is an atom. Atoms rarely exist on their own. They are like tiny balls of energy, always trying to exchange or share electrons with other atoms. They form links with one another known as bonds.

The element Carbon in diamond form

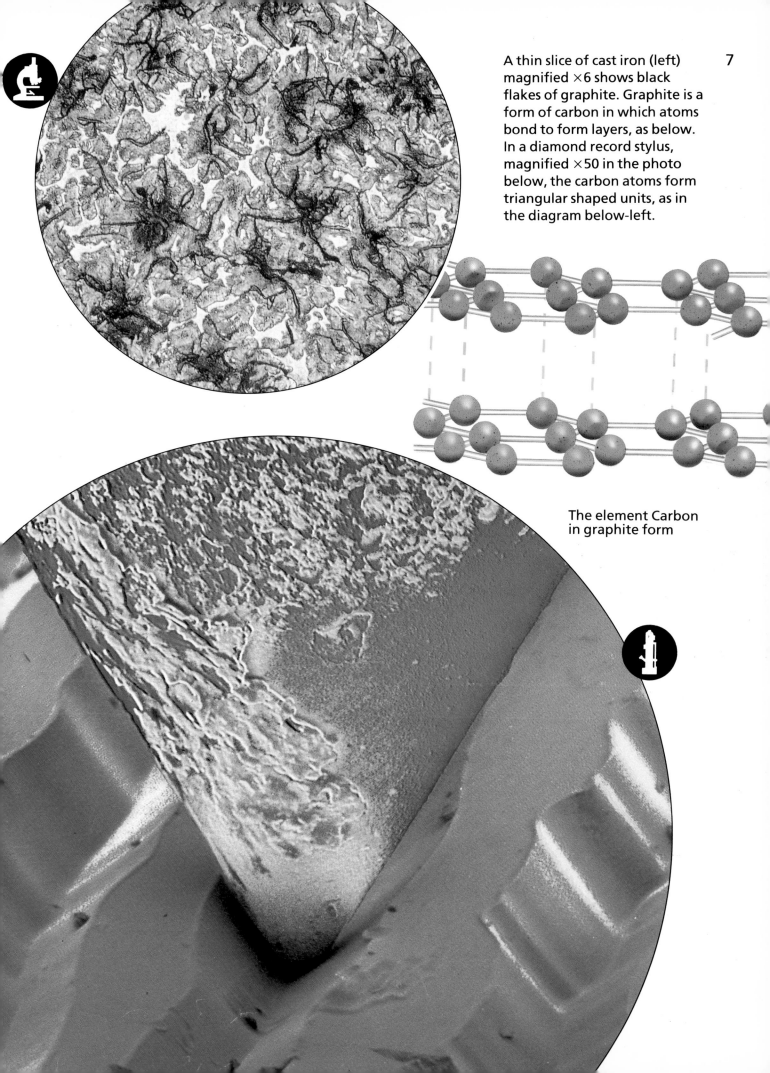

A thin slice of cast iron (left) magnified ×6 shows black flakes of graphite. Graphite is a form of carbon in which atoms bond to form layers, as below. In a diamond record stylus, magnified ×50 in the photo below, the carbon atoms form triangular shaped units, as in the diagram below-left.

The element Carbon in graphite form

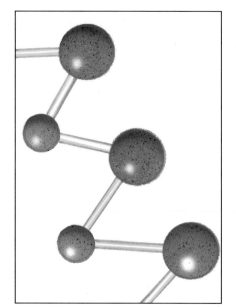

CRYSTALS AND CHAINS

Have you noticed how beautifully shaped the tiny grains of salt you have in your kitchen are? At a magnification of only ×15, as below, the grains look like minute cubes. Salt is made up of atoms of two elements, sodium and chlorine. These bond in a very regular way to form the structures we call crystals. Crystals are fascinating because they can grow even though they are not alive. Suspend a tiny crystal of copper sulphate, like one of those in the photo below-left, in a solution of the compound, and it will slowly increase in size to several inches across. Chemists – scientists who study the way that atoms bond – can make some atoms link together to form long chains of molecules.

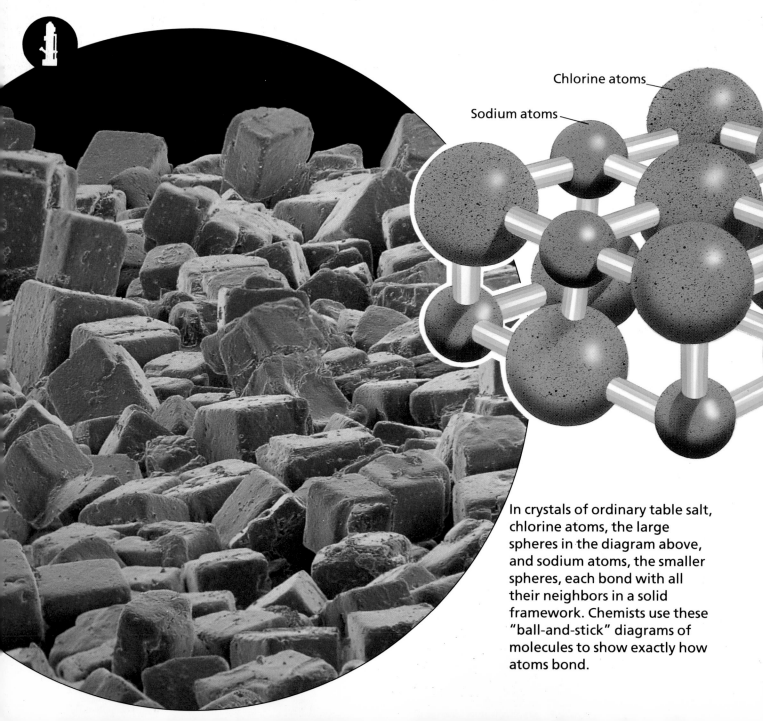

Chlorine atoms

Sodium atoms

In crystals of ordinary table salt, chlorine atoms, the large spheres in the diagram above, and sodium atoms, the smaller spheres, each bond with all their neighbors in a solid framework. Chemists use these "ball-and-stick" diagrams of molecules to show exactly how atoms bond.

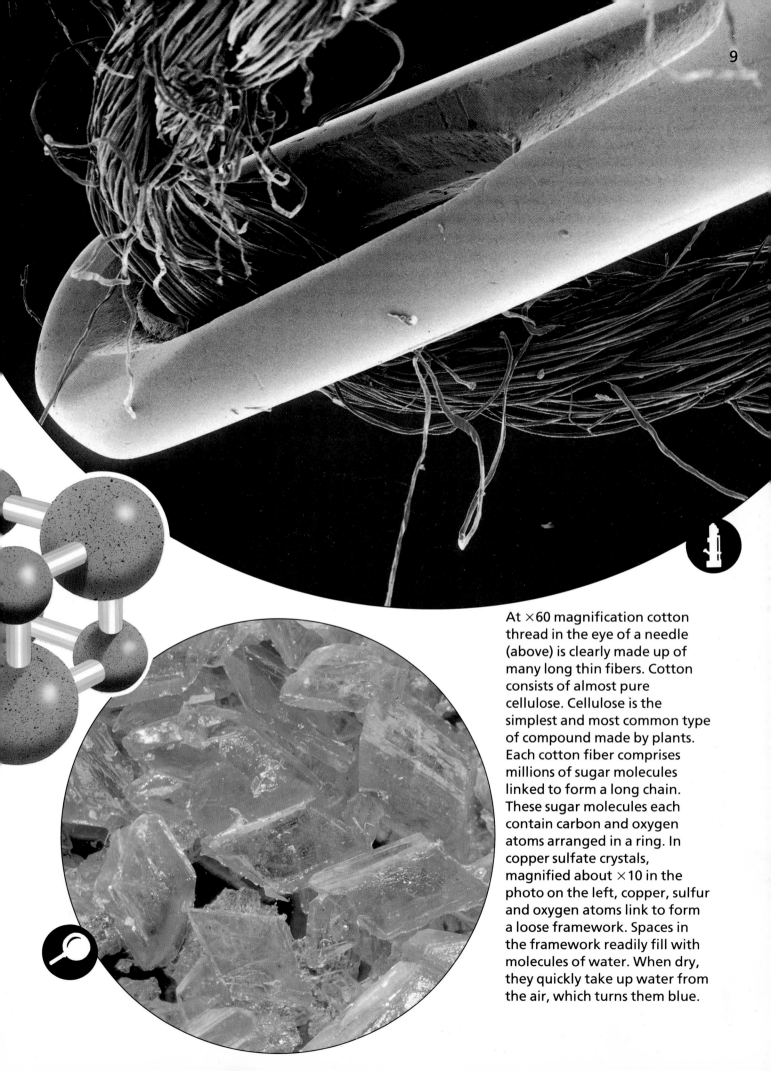

At ×60 magnification cotton thread in the eye of a needle (above) is clearly made up of many long thin fibers. Cotton consists of almost pure cellulose. Cellulose is the simplest and most common type of compound made by plants. Each cotton fiber comprises millions of sugar molecules linked to form a long chain. These sugar molecules each contain carbon and oxygen atoms arranged in a ring. In copper sulfate crystals, magnified about ×10 in the photo on the left, copper, sulfur and oxygen atoms link to form a loose framework. Spaces in the framework readily fill with molecules of water. When dry, they quickly take up water from the air, which turns them blue.

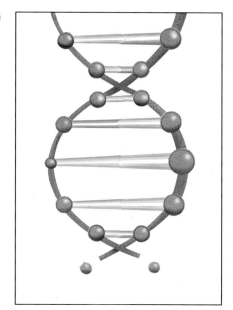

BIOMOLECULES

Carbon is one of the most important elements on Earth. All living things, including ourselves, depend on carbon compounds for growth, reproduction and repair. Those carbon compounds found in living things are called biomolecules – in Greek, *bios* means life. They include huge, but still microscopic, molecules that make up our food – fats, proteins and carbohydrates (sugars). One type of biomolecule, deoxyribonucleic acid (DNA), carries coded information that determines all the features of every animal, plant, and bacteria. Molecules of DNA are normally arranged in tiny ribbon-like structures known as chromosomes. The photos below are of human chromosomes.

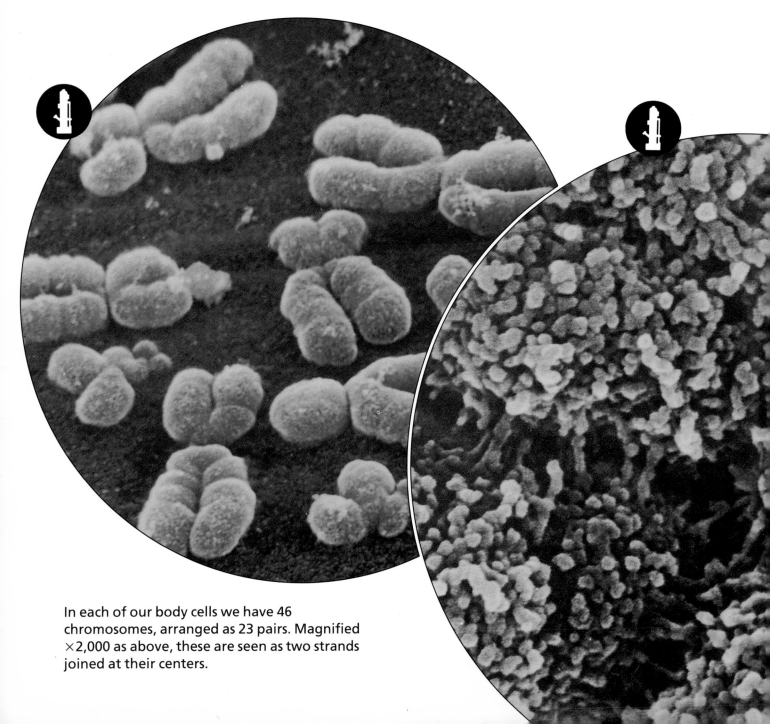

In each of our body cells we have 46 chromosomes, arranged as 23 pairs. Magnified ×2,000 as above, these are seen as two strands joined at their centers.

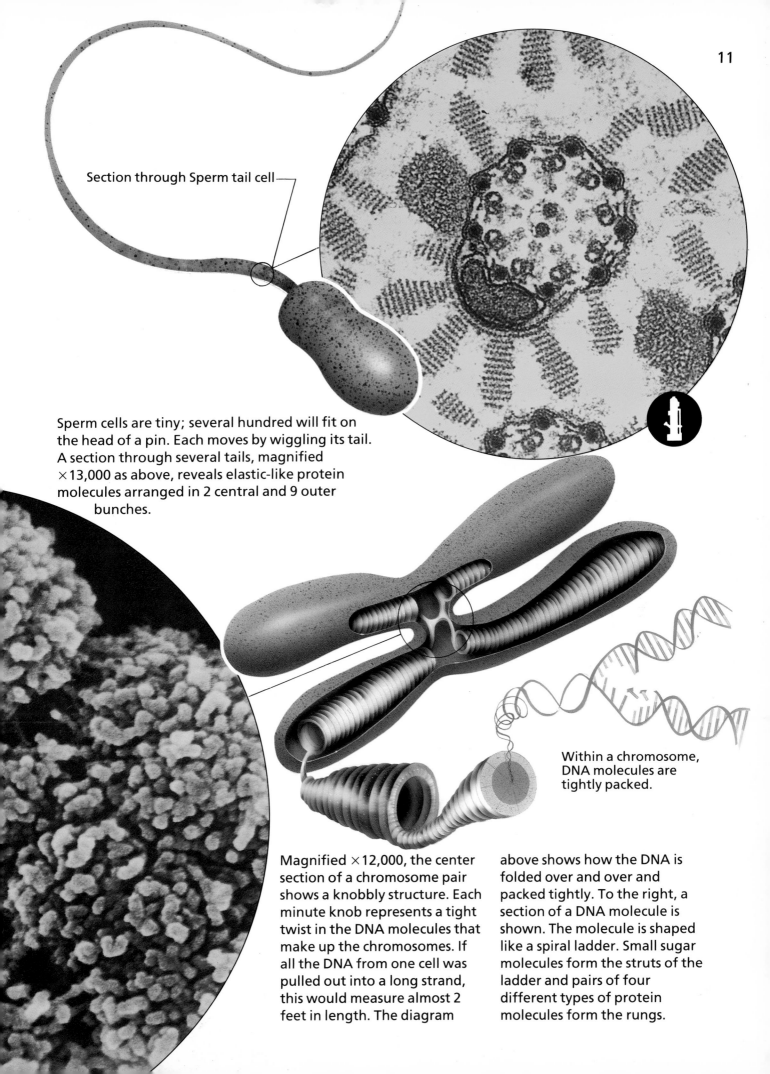

Section through Sperm tail cell

Sperm cells are tiny; several hundred will fit on the head of a pin. Each moves by wiggling its tail. A section through several tails, magnified ×13,000 as above, reveals elastic-like protein molecules arranged in 2 central and 9 outer bunches.

Within a chromosome, DNA molecules are tightly packed.

Magnified ×12,000, the center section of a chromosome pair shows a knobbly structure. Each minute knob represents a tight twist in the DNA molecules that make up the chromosomes. If all the DNA from one cell was pulled out into a long strand, this would measure almost 2 feet in length. The diagram above shows how the DNA is folded over and over and packed tightly. To the right, a section of a DNA molecule is shown. The molecule is shaped like a spiral ladder. Small sugar molecules form the struts of the ladder and pairs of four different types of protein molecules form the rungs.

VIRUSES

We say something is "living" if it can grow, reproduce and react to changes in its surroundings. A virus, by itself, can do none of these things. But once inside a live cell, a virus takes over that cell and turns it into thousands of new viruses. These escape from the dead cell and invade other cells. If the cells try to fight back, the virus can often change its shape and form to overcome their defenses. So viruses are on the borderline between living and non-living. They are in fact incredibly small packages of just two main types of biomolecules, proteins and nucleic acids (see pages 10-11). All viruses are too small to be seen with a home microscope, but electron microscope photos, as shown here, reveal them.

The photo above is of a human white blood cell infected with the AIDS virus, magnified several thousand times. The image has been colored using a computer.

The cell has a lumpy appearance due to the mass of viruses growing inside it. The tiny green blobs on the surface of the cell are AIDS viruses about to burst free.

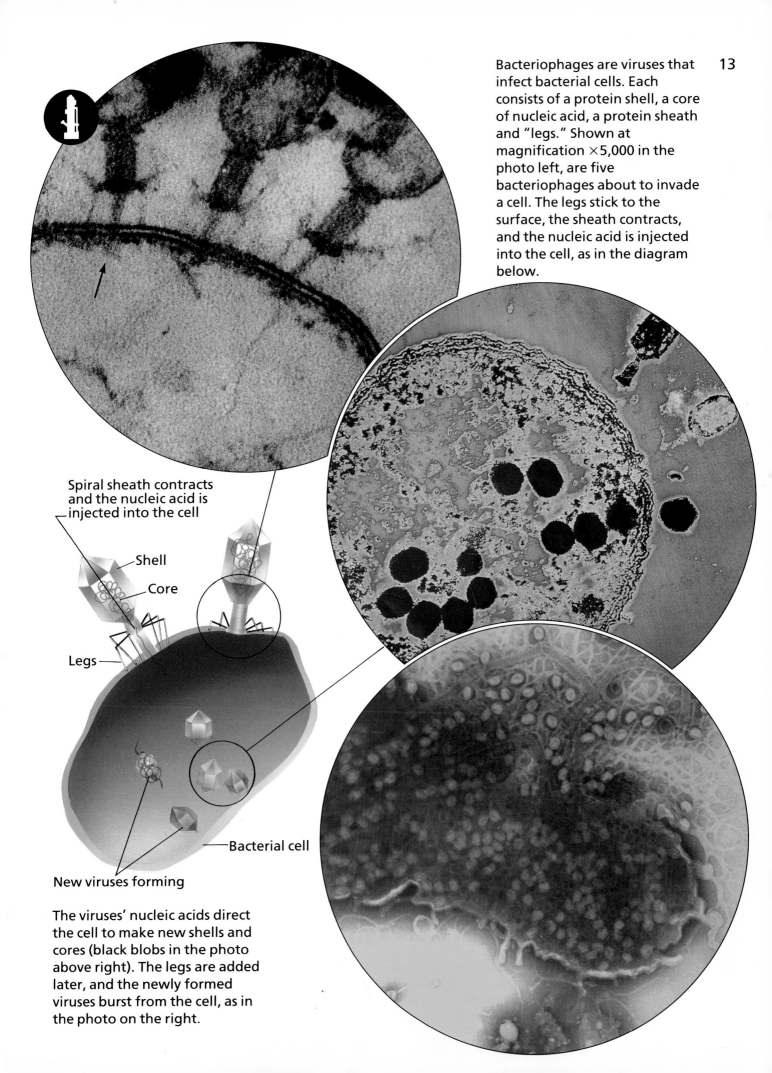

Bacteriophages are viruses that infect bacterial cells. Each consists of a protein shell, a core of nucleic acid, a protein sheath and "legs." Shown at magnification ×5,000 in the photo left, are five bacteriophages about to invade a cell. The legs stick to the surface, the sheath contracts, and the nucleic acid is injected into the cell, as in the diagram below.

Spiral sheath contracts and the nucleic acid is injected into the cell

Shell

Core

Legs

Bacterial cell

New viruses forming

The viruses' nucleic acids direct the cell to make new shells and cores (black blobs in the photo above right). The legs are added later, and the newly formed viruses burst from the cell, as in the photo on the right.

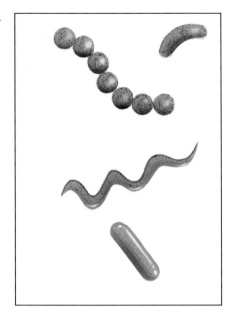

BACTERIA

Bacteria are tiny creatures that we often call germs. Indeed, most of us think of them as living things that cause nasty diseases. Bacteria are responsible for such human diseases as food poisoning and pneumonia. But many bacteria are extremely helpful to us. In the soil, some break down dead animals and plants and make their chemicals available to other organisms. Bacteria are also responsible for turning milk into yoghurt, cheese and butter. All bacteria are about 10,000 times larger than viruses but even the biggest of them can only just be seen with a home microscope. The photo below, at a magnification of more than ×1,000, shows that many thousands of bacteria can fit on the head of a pin.

Bacteria come in all shapes and forms. Some are simply round or oval, as in the photos below and on the right. Others are rod, comma, club or even spiral-shaped, as seen below left. The bacteria shown below live in water and cause Legionnaire's disease.

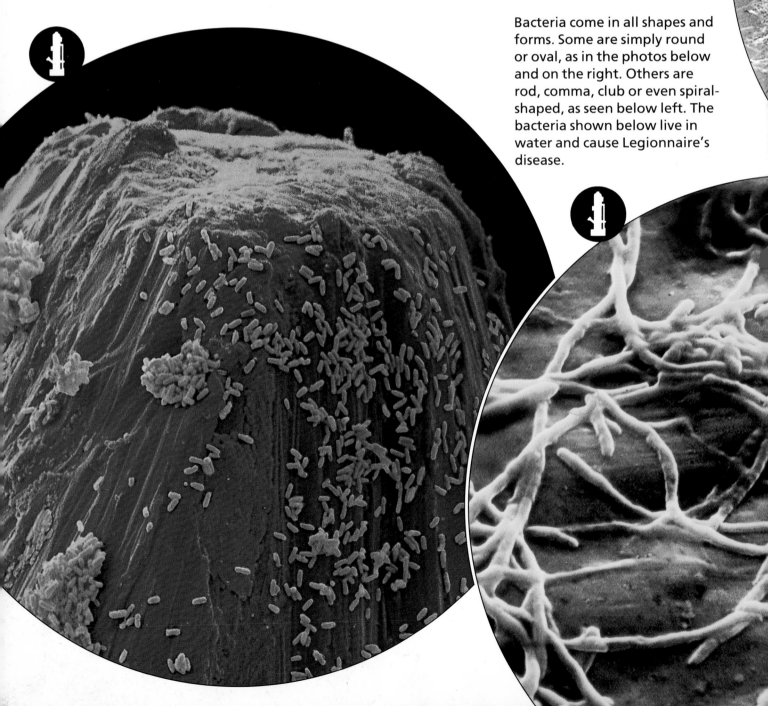

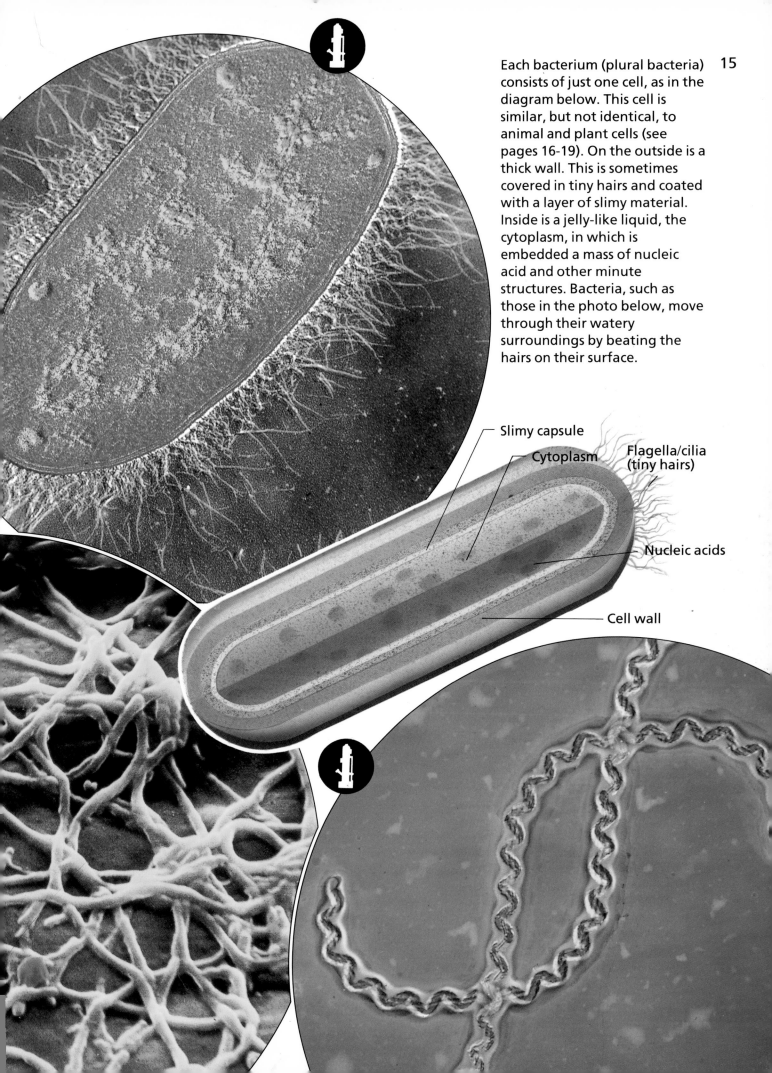

Each bacterium (plural bacteria) consists of just one cell, as in the diagram below. This cell is similar, but not identical, to animal and plant cells (see pages 16-19). On the outside is a thick wall. This is sometimes covered in tiny hairs and coated with a layer of slimy material. Inside is a jelly-like liquid, the cytoplasm, in which is embedded a mass of nucleic acid and other minute structures. Bacteria, such as those in the photo below, move through their watery surroundings by beating the hairs on their surface.

Slimy capsule

Cytoplasm

Flagella/cilia (tiny hairs)

Nucleic acids

Cell wall

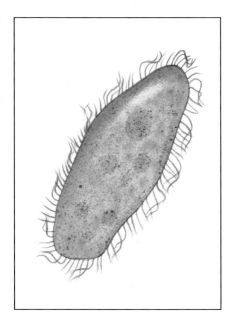

SINGLE-CELL ANIMALS

If you look at a drop of pond or sea water under a home microscope or even just a hand-lens, you are likely to see a number of tiny animals moving about. Some of these, like *Amoeba* and *Paramecium* photographed below, are single-cell creatures. A typical animal cell comprises a jelly-like outer membrane, the ectoplasm, and an inner liquid material, the endoplasm (or cytoplasm). Floating around in the cell are various structures such as the nucleus, the cell's control center. A contractile vacuole is responsible for collecting unwanted water in the cell and then, when full, bursting to get rid of its contents to the surroundings. Each single-cell animal is capable of feeding, growing and reproducing on its own.

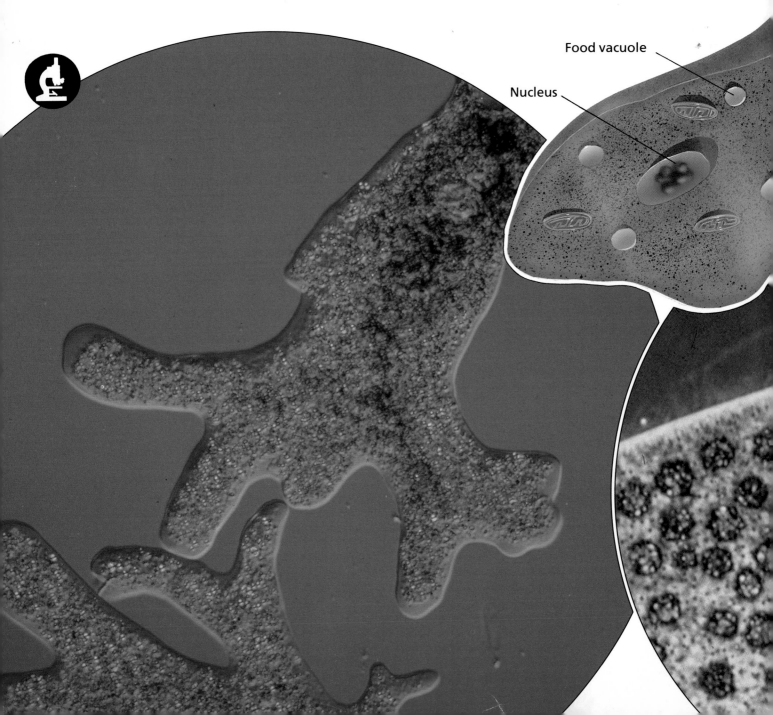

Food vacuole

Nucleus

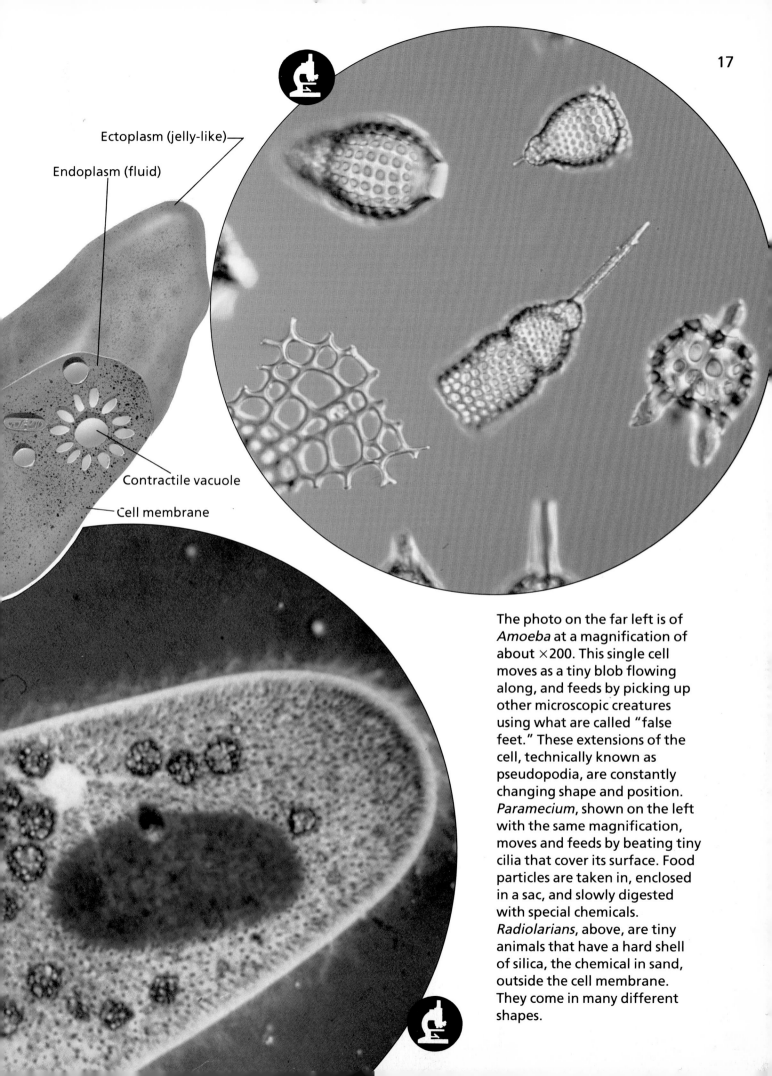

Ectoplasm (jelly-like)

Endoplasm (fluid)

Contractile vacuole

Cell membrane

The photo on the far left is of *Amoeba* at a magnification of about ×200. This single cell moves as a tiny blob flowing along, and feeds by picking up other microscopic creatures using what are called "false feet." These extensions of the cell, technically known as pseudopodia, are constantly changing shape and position. *Paramecium*, shown on the left with the same magnification, moves and feeds by beating tiny cilia that cover its surface. Food particles are taken in, enclosed in a sac, and slowly digested with special chemicals. *Radiolarians*, above, are tiny animals that have a hard shell of silica, the chemical in sand, outside the cell membrane. They come in many different shapes.

SINGLE-CELL PLANTS

A sample of pond water (see page 16) is also likely to contain a variety of single-cell plants. Plants differ from animals in two main ways. First, plants can make their own food and second, because of this, plants do not need to move about very much, if at all. Animal cells feed on plant and animal material and break down the large complex chemicals within this to get essential nutrients and energy. Plant cells use the energy of sunlight to convert carbon dioxide and water into sugars. This is known as photosynthesis. They then convert these sugars into other nutrients or break them down to release energy. All the microscopic single-cell plants shown below and opposite are capable of photosynthesis.

Within each plant cell are special structures known as chloroplasts. These contain the green pigment chlorophyll which traps the energy of sunlight to make sugars. Around each plant cell is a thick wall made of cellulose (see page 9). This gives the cell a rigid structure and protects the soft structure inside. The diagram below right shows a section through a typical plant cell.

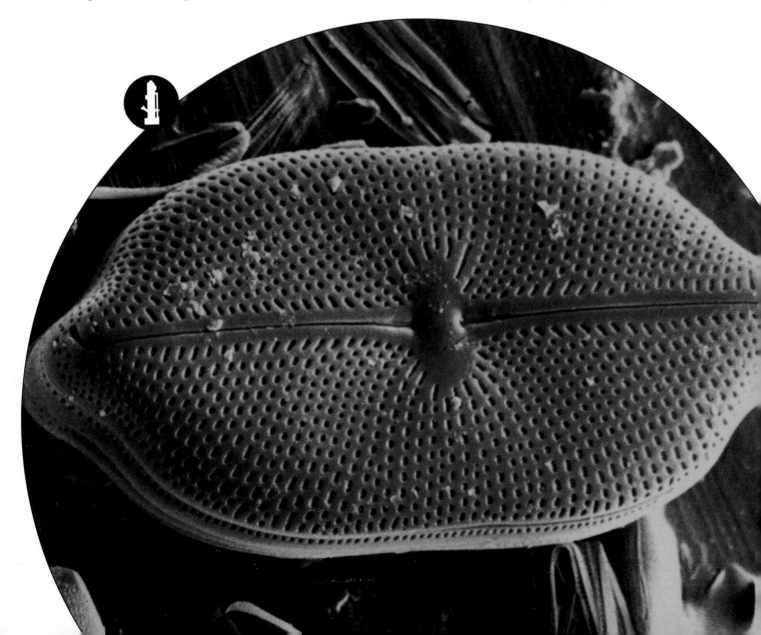

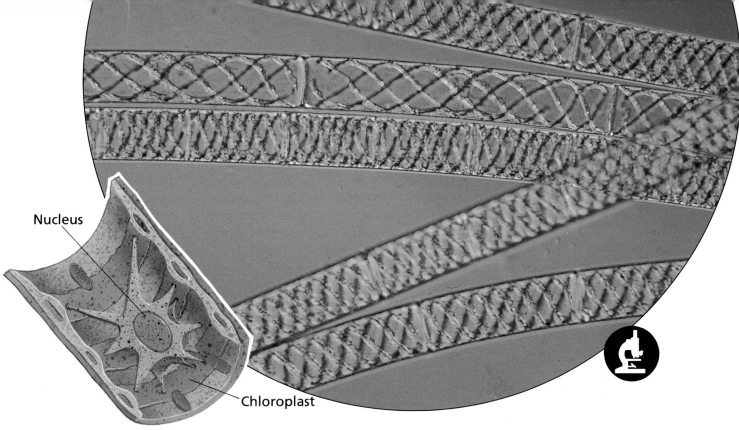

Nucleus

Chloroplast

The most common single-cell plants are types of algae. Algae are found in ponds, rivers and seas throughout the world. They range in size from microscopic *diatoms* and *Chlamydomonas* to giant, many-celled seaweeds 60m (200ft) long. *Diatom* cells, like the one shown below left at about 3,000 times life-size, have a shell-like case of silica. The case is made up of two equal halves. *Spirogyra*, above, is a common pond algae in which many single cells are joined end to end to form long thin slimy threads or filaments. The filaments can be seen with the naked eye. At a magnification of ×100, as in the photo above, the green chloroplasts are visible as spiral bands within each cell. *Chlamydomonas*, at ×200 life-size as below, swims by beating flagella that stick out at one end. They can both photosynthesize and move about quite well.

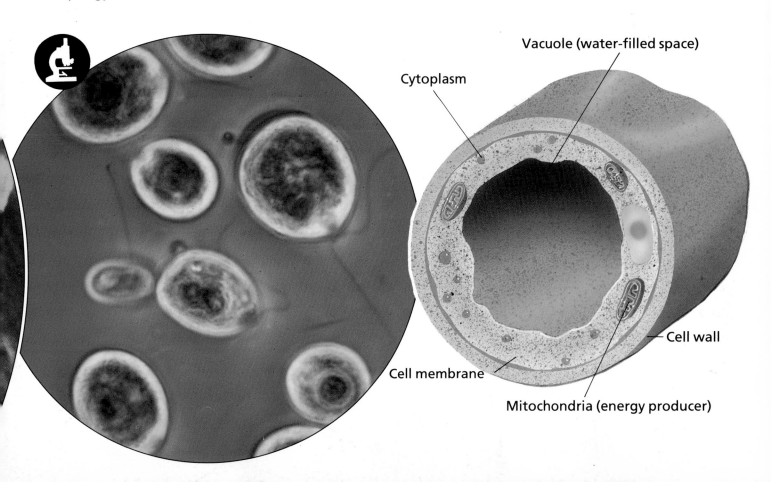

Vacuole (water-filled space)

Cytoplasm

Cell wall

Cell membrane

Mitochondria (energy producer)

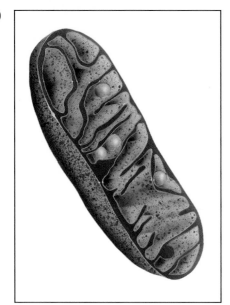

ORGANELLES

Cells are the building blocks of living things and the smallest units of life that can exist on their own. Within cells there are several minute structures called organelles, each of which has a specific job to do. Large cells, such as the human egg cell, or ovum, measure about 0.004 inches (0.1mm) across and can be seen with the naked eye. But only under a high power microscope can the organelles within them be seen. The largest of these is the nucleus. This directs all the cell's activities. Sausage-shaped organelles known as *mitochondria* are the main sites of energy production. Within many-celled creatures, each cell is adapted for a certain job. Some of these cells have organelles not shared by other cells.

The sausage-shaped organelles illustrated on the top left and right of these pages, are known as *mitochondria*. Only visible through a high-powered microscope, we can see them end-on in the picture below-left and also below-right.

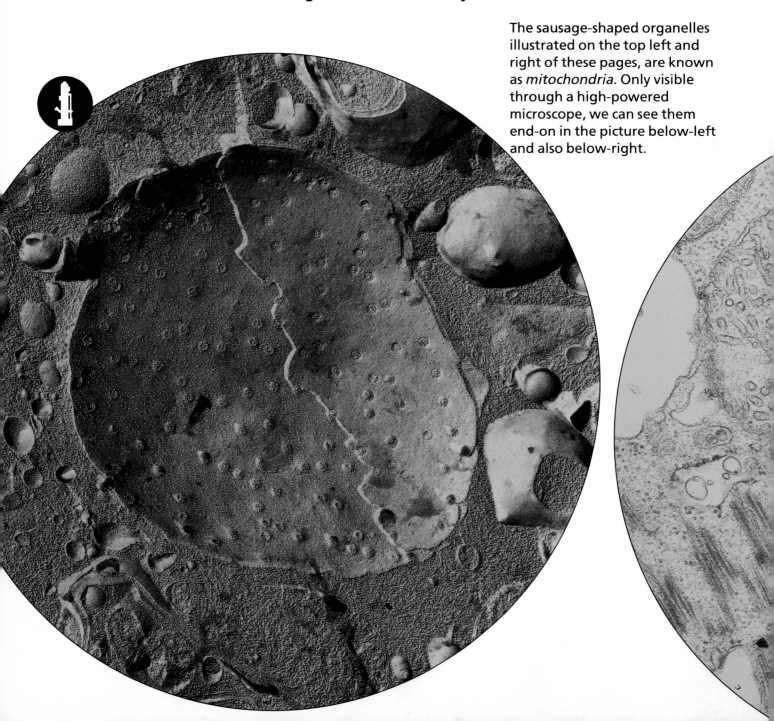

The cell nucleus, seen magnified more than 15,000 times in the photo below left, is surrounded by a thin membrane or covering which has many tiny holes in it. This envelope acts as a sieve, controlling the constant flow of chemicals to and from the nucleus. Surrounding each cell is another envelope, or membrane, which acts as a control gate for the entry of nutrients and the exit of wastes from the cell. In a cell from the human nose magnified ×7,000 (photo below), many minute hairs stick out from the outer membrane. These special organelles can beat from side to side to remove dust particles from the air a person breathes.

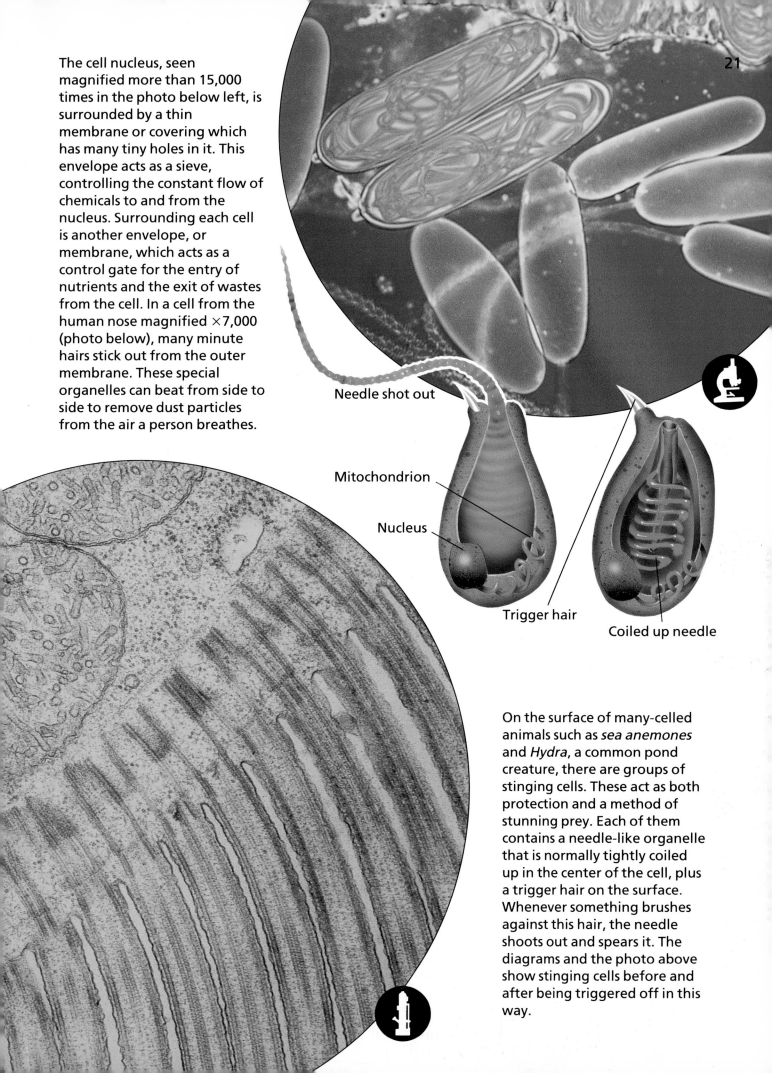

Needle shot out

Mitochondrion

Nucleus

Trigger hair

Coiled up needle

On the surface of many-celled animals such as *sea anemones* and *Hydra*, a common pond creature, there are groups of stinging cells. These act as both protection and a method of stunning prey. Each of them contains a needle-like organelle that is normally tightly coiled up in the center of the cell, plus a trigger hair on the surface. Whenever something brushes against this hair, the needle shoots out and spears it. The diagrams and the photo above show stinging cells before and after being triggered off in this way.

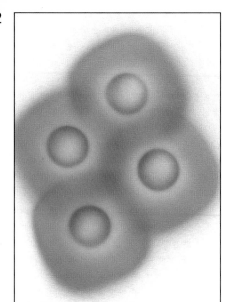

CELL DIVISION

All living things start life as a single cell. The cell grows and eventually it may split into two equal halves. In turn, each of these daughter cells divides into two. The root of a plant such as an onion gets longer and longer as cells within it – seen below at about 100 times life-size – divide repeatedly every few days. As babies, we all grow in size as many thousands of individual cells in our body each divide, until there are some hundreds of millions of cells. In order that one cell can become two, all the organelles inside must be copied and shared out evenly. This happens in four distinct stages, as shown in the diagrams below. Here, the two new cells have exactly the same number and type of chromosomes (see pages 10-11).

Prophase

Centrioles, to which the chromosomes attach

Nuclear membrane

Chromosome pairs

Metaphase

Anaphase

Telophase

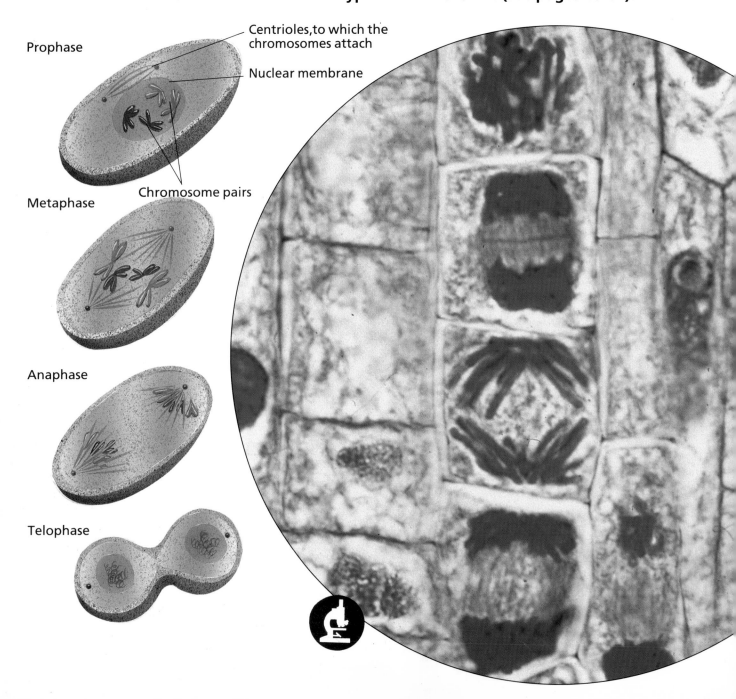

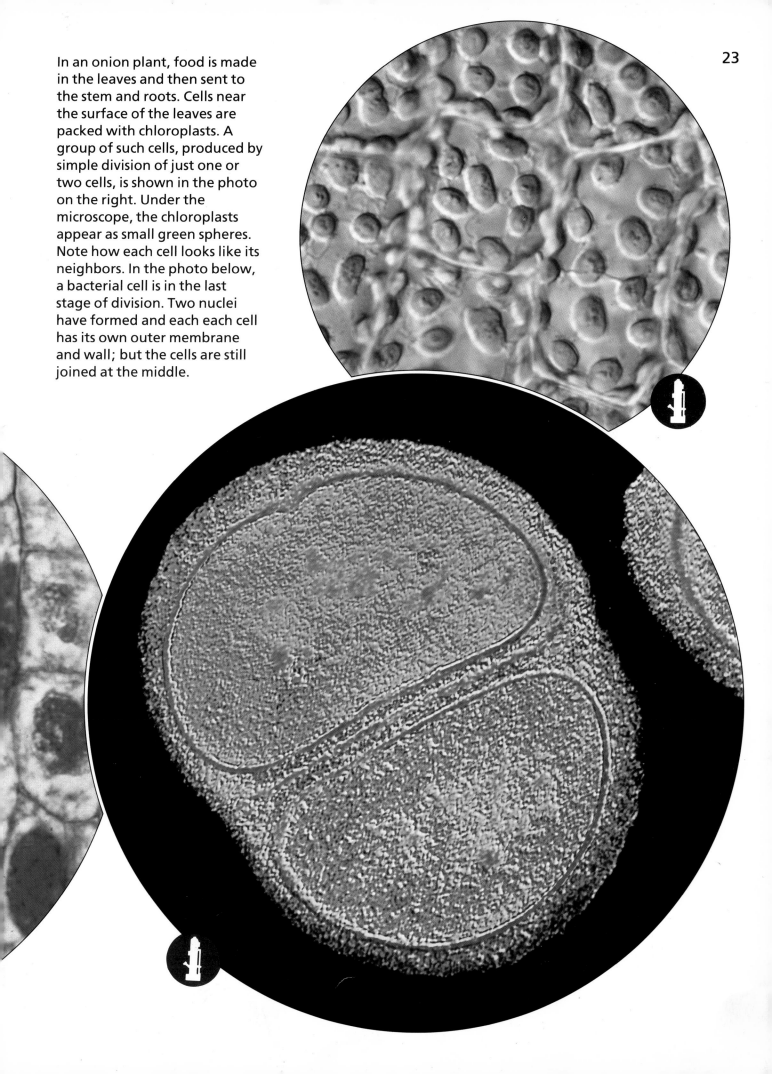

In an onion plant, food is made in the leaves and then sent to the stem and roots. Cells near the surface of the leaves are packed with chloroplasts. A group of such cells, produced by simple division of just one or two cells, is shown in the photo on the right. Under the microscope, the chloroplasts appear as small green spheres. Note how each cell looks like its neighbors. In the photo below, a bacterial cell is in the last stage of division. Two nuclei have formed and each each cell has its own outer membrane and wall; but the cells are still joined at the middle.

SPORES

The type of cell division we looked at on pages 22-23 is how all creatures increase in size and how some, for example bacteria and Amoeba, multiply. However, flowering plants and animals such as ourselves multiply by producing special sex cells – pollen/sperm and eggs – which, when they combine, give rise to a new individual. Fungi, ferns and algae, in a similar way, form special cells for producing offspring. These cells are known as spores. Spores are microscopic structures which are blown by the wind or carried by water currents away from the parent creature. In this way they help to avoid overcrowding of the organisms. When conditions are suitable, the spore cells grow and divide to produce a new individual.

Fungi such as the common bread mold consist of a network of fine thread-like feeding structures. They are known as hyphae. At intervals along their length, hyphae form reproductive units called sporangia, as seen in the photo below. Each sporangium contains thousands of spores.

A common woodland fern, shown right, produces spores in capsules on the underside of its fronds (leaves). As the capsules dry out, they split and throw out the spores. In the photo far right, three spore capsules, at different stages of splitting and shedding their contents, are shown at about ×200 life-size.

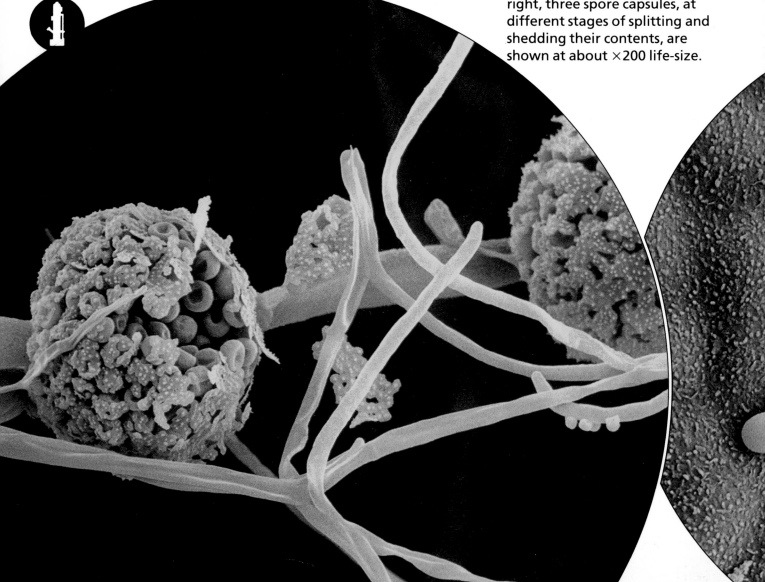

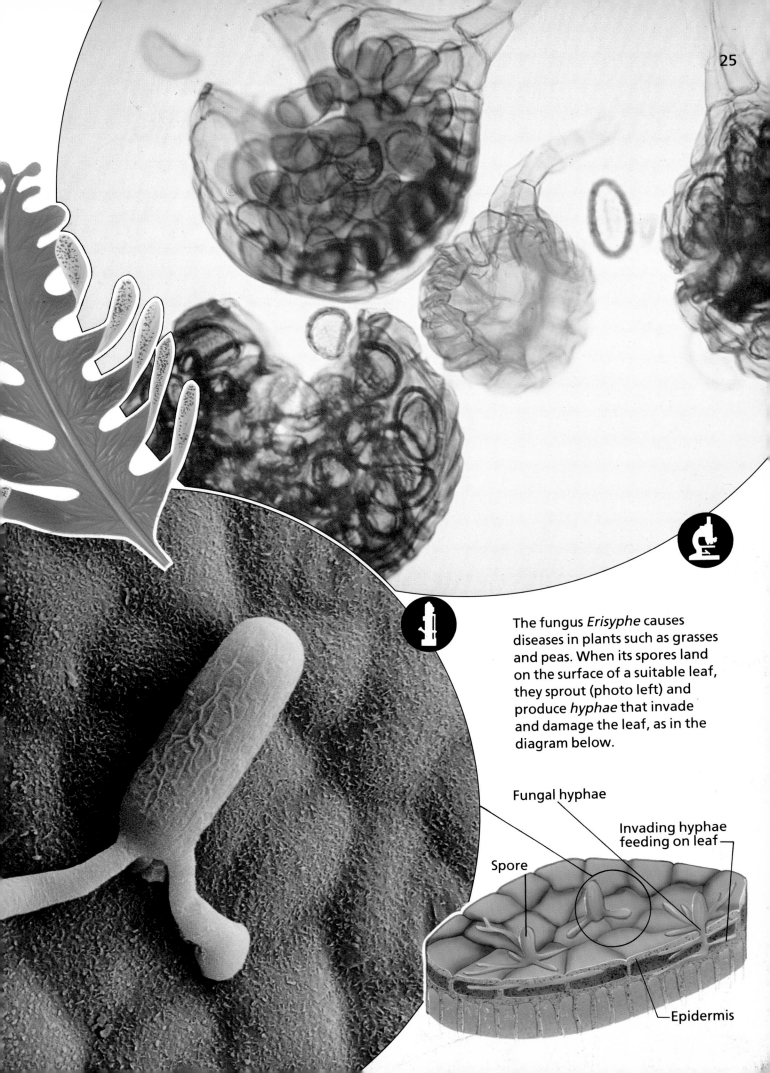

The fungus *Erisyphe* causes diseases in plants such as grasses and peas. When its spores land on the surface of a suitable leaf, they sprout (photo left) and produce *hyphae* that invade and damage the leaf, as in the diagram below.

Fungal hyphae

Invading hyphae feeding on leaf

Spore

Epidermis

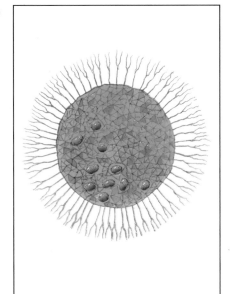

SIMPLE MANY-CELLED CREATURES

The alga *Spirogyra* that we looked at on pages 18-19, is made up of many cells, each independent of its neighbors. In another simple creature, *Volvox*, shown below at a magnification of about ×200, thousands of almost identical cells are joined together to form a hollow ball of cells. In this so-called "colony," the cells do interact with, and depend on, one another. The jellyfish and the water flea, *Daphnia*, shown opposite, also have a many-celled structure, although in these, different types of cells are present. These are arranged in groups called tissues, and groups of tissues form organs.

Volvox often reproduces by certain cells with the colony each dividing several times to form small daughter colonies. A few of these are visible near the center of the photo below. They break free from the parent colony by splitting it open.

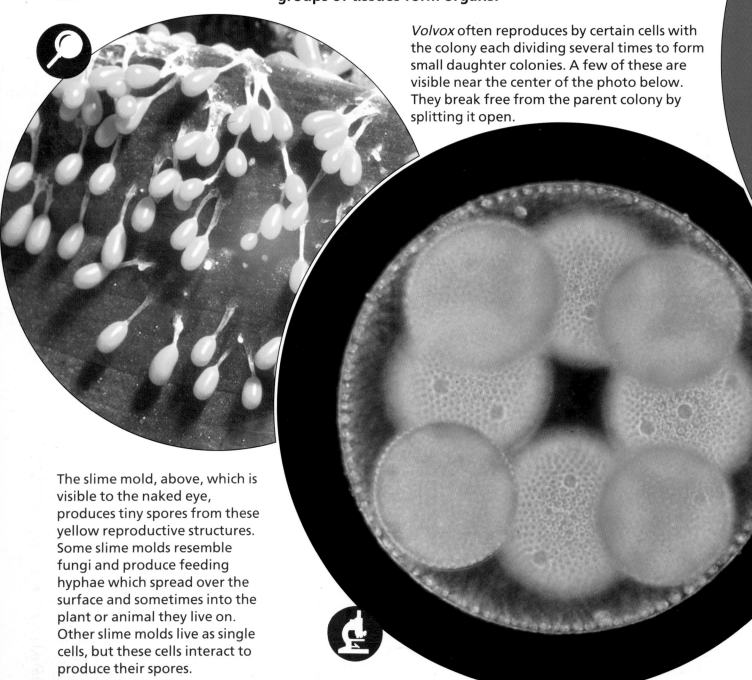

The slime mold, above, which is visible to the naked eye, produces tiny spores from these yellow reproductive structures. Some slime molds resemble fungi and produce feeding hyphae which spread over the surface and sometimes into the plant or animal they live on. Other slime molds live as single cells, but these cells interact to produce their spores.

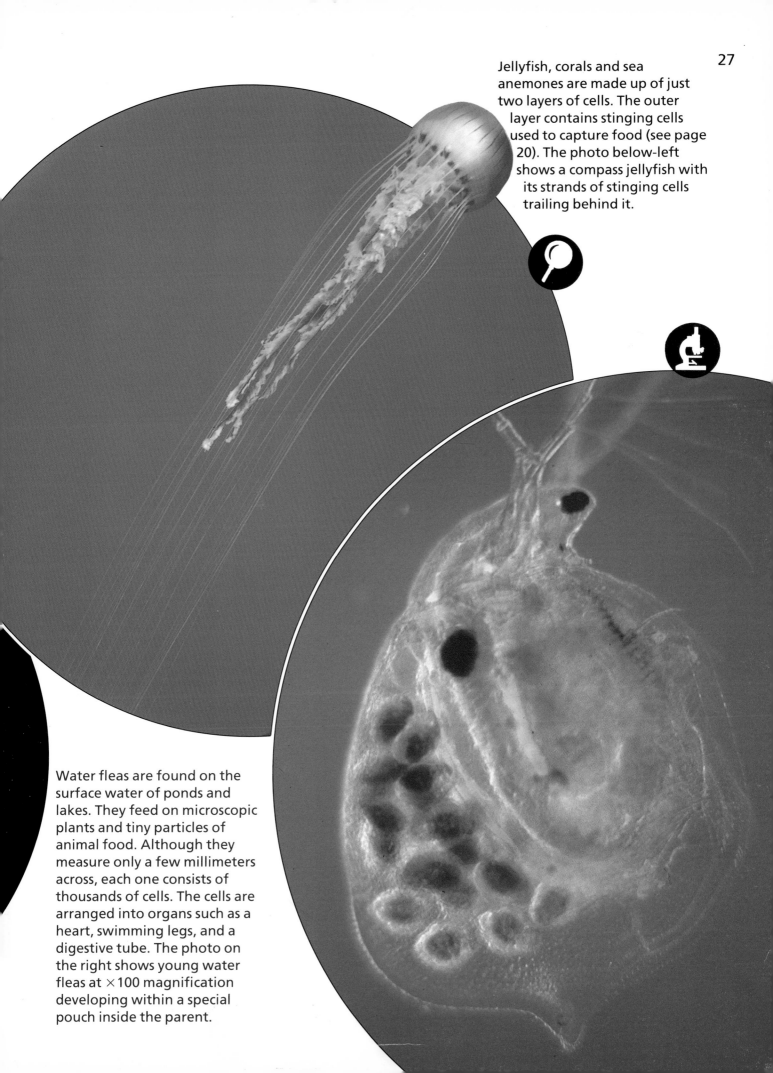

Jellyfish, corals and sea anemones are made up of just two layers of cells. The outer layer contains stinging cells used to capture food (see page 20). The photo below-left shows a compass jellyfish with its strands of stinging cells trailing behind it.

Water fleas are found on the surface water of ponds and lakes. They feed on microscopic plants and tiny particles of animal food. Although they measure only a few millimeters across, each one consists of thousands of cells. The cells are arranged into organs such as a heart, swimming legs, and a digestive tube. The photo on the right shows young water fleas at ×100 magnification developing within a special pouch inside the parent.

PRACTICAL PROJECTS

You can discover a great deal about Nature's miniature world with just a hand-lens. But to see greater detail you will need a home microscope like that shown on page 4. The objects you wish to study must be mounted on a glass slide. They must be made so that light can shine through them, so you may need to cut very thin slices of plant and animal material. To pick out the different types of cell, and the structures within them, you will need to stain your specimens. The way to do this is outlined below. If you are going to try something tricky, it is worth getting help from an adult. You may be able to start your studies with some ready-made slides bought from a microscope supplier.

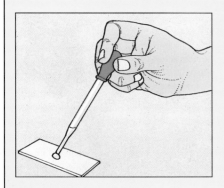

To prepare a slide of cells, place a drop of clean water containing them on the glass.

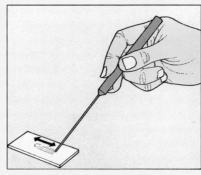

With a wire loop that has been sterilized in a flame, spread the fluid thinly.

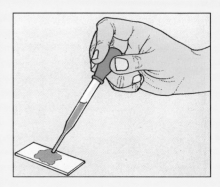

Add a small drop of staining dye to the cells and leave for a few minutes.

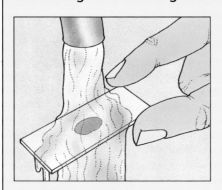

Wash off the dye with water or alcohol. You can stain with another, contrasting dye.

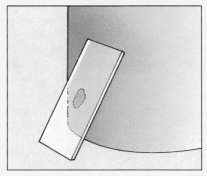

Leave the slide to dry. You can speed up drying by warming the slide over a flame.

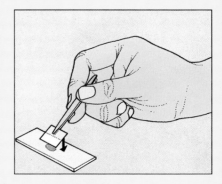

Place the cover slip (a thin square of glass) over the stained cells.

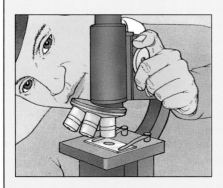

Put the slide on the microscope stage and position the mirror to give you good illumination.

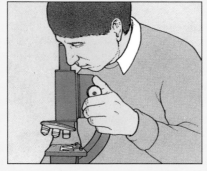

Select the objective lens you want, then move the eyepiece up or down to focus.

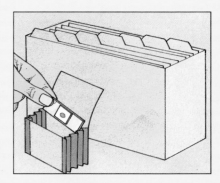

Keep your prepared slides in a cardboard wallet made by folding a thin sheet of card.

29

Bacteria are best studied using a home microscope. The cells should be spread out onto a slide and stained with a special dye called "Gram's stain."

With a hand-lens you can study the structure and shape of fungi or of slime molds, such as these on the surface of the bark of a tree.

A sample of water from a stagnant pond should contain a mass of tiny animals and plants that you can study with a hand-lens or a home microscope (above). Details of each creature, such as the nucleus and food vacuoles, can often be seen without having to stain the cells. To watch the tiny animals move about, you will probably have to keep focusing and refocusing as they swim in and out of view.

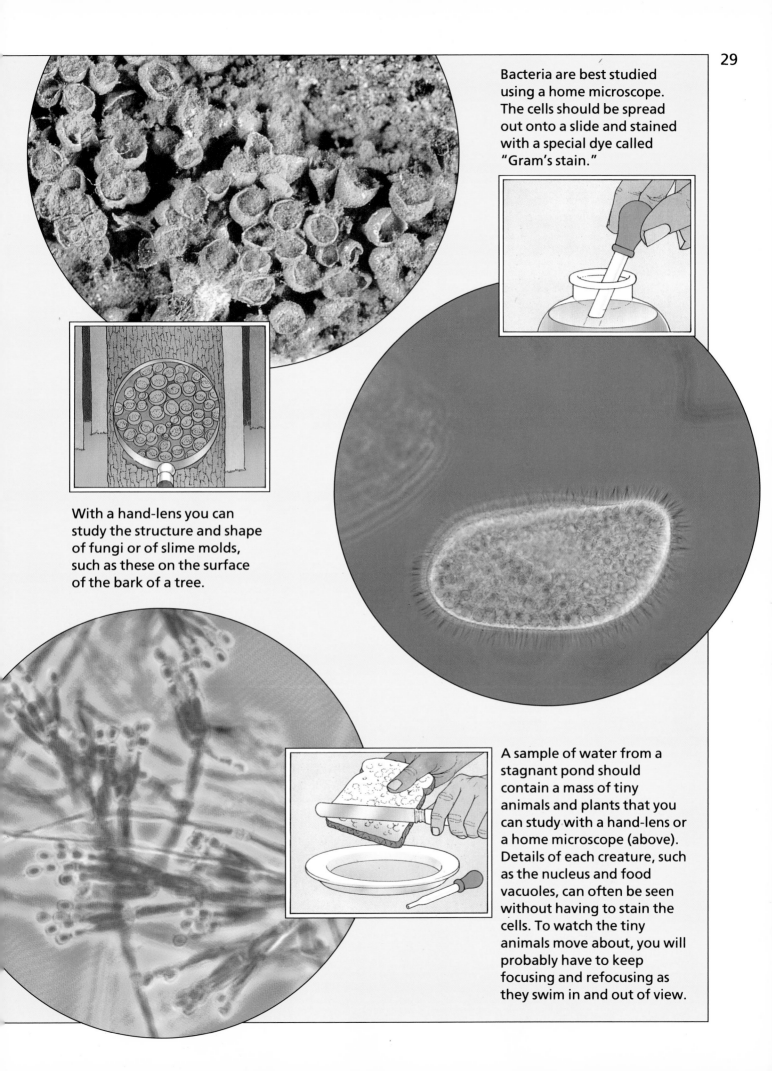

MICROPHOTOGRAPHY

With ordinary scientific instruments such as fine tweezers and liquid-droppers (pipettes), it is possible to handle and manipulate individual cells such as those of *Amoeba*. However, to handle the tiny structures within cells, the organelles, especially small and fine (micro) instruments and a high-powered binocular microscope are needed. Such a microscope allows minute objects be seen in three-dimensions, so giving the image "depth." By looking at a cell under such a high-powered binocular microscope and using a micropipette, the nucleus of a cell can be removed and then mounted on a slide for further examination by a skilled technician in the laboratory.

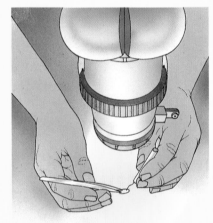

To prepare a slide for an electron microscope the organelle must be handled with a micro-pipette very delicately. Once it has been sliced it is sprayed with gold. This reflects the electron beam or creates a shadow of the image so that the image can appear clearly under the electron microscope.

There are two main types of electron microscope. In a transmission type, a beam of electrons is passed through an extremely thin slice of tissue and an image is produced on a viewing screen. On a scanning electron microscope (an SEM), a fine beam of electrons is moved across the surface of the tissue for reflections to be collected and used to create an image on a television type of screen. Using an SEM, realistic 3-D images are produced. But as with all types of microscope specimens, the tissues and organs are no longer alive. The slide preparation process kills live cells. The colors on photos produced using an SEM are false colors added in processing.

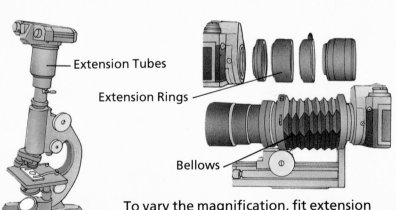

Extension Tubes

Extension Rings

Bellows

To vary the magnification, fit extension rings or bellows to the camera attachment.

GLOSSARY

atom the smallest particle that can exist. Normally, atoms are indivisible. All matter — solids, liquids and gases — are made up of atoms.

bacteria small single-cell organisms without chlorophyll but with a cell wall and simple nucleus. They reproduce quickly by splitting in two equal parts.

biomolecule a type of molecule found in living things, such as sugars, fats and proteins.

capsule a structure with a tough wall, such as that containing the spores of fungi and ferns.

cell one of the building blocks of which living things are made. Some creatures consist of just one cell, others of many thousands or millions. Some cells may be specialized for particular jobs.

chlorophyll the green pigment of plants and some simple single-cell creatures that absorbs the energy of sunlight so it can be used to produce sugars.

compound a substance made up of a mixture of atoms, such as salt, which is made up of atoms of sodium and chlorine.

crystal a regularly shaped structure created by the precise way in which some atoms combine together in elements and compounds. Diamond is a crystal formed by the regular bonds between atoms of carbon. Salt grains get their cube-shaped form from the regular links between sodium and chlorine atoms.

DNA short for deoxyribonucleic acid. DNA is a chemical blueprint for living things. It carries coded information that determines the features of a creature and ensures that these are copied and passed on to any offspring.

element a substance composed of just one type of atom, such as gold, silver, oxygen gas.

frond the large divided leaf of a fern, or the flattened body of a seaweed.

hyphae the branching filaments that make up the body of many fungi and slime molds.

magnification the number of times larger that an object seen through a lens or microscope appears compared with its true size.

molecule a combination of two or more atoms. A molecule of water, for example, consists of one atom of oxygen combined with two atoms of hydrogen.

organelle a distinct structure within a living cell, such as the nucleus.

pigment a substance that is colored. The green of plants is caused by the pigment chlorophyll.

spore a resting or dispersal stage of many simple creatures, for example most single-cell animals and plants, bacteria, fungi and ferns.

virus a tiny organism, too small to be seen with even a home microscope, which can only multiply inside a living cell. Viruses cause unwelcome changes — diseases — in the "host" cells and organisms they infect.

WEIGHTS AND MEASURES

mm = millimeter 10mm = 1cm = 4/10 inch
cm = centimeter 100cm = 1m = 3 1/3 feet
m = meter 1,000m = 1km = 6/10 mile
km = kilometer   0.1 = 1/10
lb = pound g = gram 1,000g = 1kg = 2lb 3 ounces   0.01 = 1/100
kg = kilogram   0.001 = 1/1,000

INDEX

Photographic Credits:
Cover and pages 6, 7 both, 8, 10, 10-11, 11, 12, 13 all, 14, 15, 16, 17 both, 19t, 21 both, 22, 23b, 24, 25 both, 26 both, 27r and 29m: Science Photo Library; pages 9 both, 18, 19b, 20, 23t, 29t and b, and 30: Biophoto Associates; page 27l: Pete Atkinson/Planet Earth.

PRINTED IN BELGIUM BY

INTERNATIONAL BOOK PRODUCTION